LET'S EXPLORE THE WEATHER!

By Nicole Horning

Cavendish Square

New York

Published in 2021 by Cavendish Square Publishing, LLC
243 5th Avenue, Suite 136, New York, NY 10016

First Edition

Website: cavendishsq.com

This publication represents the opinions and views of the author based on his or her personal experience, knowledge, and research. The information in this book serves as a general guide only. The author and publisher have used their best efforts in preparing this book and disclaim liability rising directly or indirectly from the use and application of this book.

All websites were available and accurate when this book was sent to press.

Library of Congress Cataloging-in-Publication Data

Names: Horning, Nicole, author.
Title: Let's explore the weather! / Nicole Horning.
Description: New York : Cavendish Square Publishing, [2021] |
Series: Earth science explorers | Includes index.
Identifiers: LCCN 2019049050 (print) | LCCN 2019049051 (ebook) |
ISBN 9781502656438 (library binding) | ISBN 9781502656414 (paperback) |
ISBN 9781502656421 (set) | ISBN 9781502656445 (ebook)
Subjects: LCSH: Weather–Juvenile literature.
Classification: LCC QC981.3 .H675 2021 (print) | LCC QC981.3 (ebook) |
DDC 551.5–dc23
LC record available at https://lccn.loc.gov/2019049050
LC ebook record available at https://lccn.loc.gov/2019049051

Editor: Nicole Horning
Copy Editor: Nathan Heidelberger
Designer: Rachel Rising

The photographs in this book are used by permission and through the courtesy of: Cover Sasa Prudkov/Shutterstock.com; p. 5 New Africa/Shutterstock.com; p. 7 liseykina/Shutterstock.com; p. 9 kostasgr/Shutterstock.com; p. 11 K Stocker/Shutterstock.com; p. 13 irin-k/Shutterstock.com; p. 15 (top) February_Love/Shutterstock.com; p. 15 (middle) Datskevich Aleh/Shutterstock.com; p. 15 (bottom) Pinglabel/Shutterstock.com; 17 Roman Mikhailiuk/Shutterstock.com; p. 19 Suzanne Tucker/Shutterstock.com; p. 21 Delbars/Shutterstock.com; p. 23 gorillaimages/Shutterstock.com.

Some of the images in this book illustrate individuals who are models. The depictions do not imply actual situations or events.

CPSIA compliance information: Batch #CS20CSQ: For further information contact Cavendish Square Publishing LLC, New York, New York, at 1-877-980-4450.

Printed in the United States of America

Find us on

CONTENTS

What Is Weather?

Weather is what's happening in the air around us. Rain, snow, and wind are all kinds of weather. The weather also includes how hot or cold it is. People wear different clothes in different kinds of weather!

5

The sun plays a big part in the weather. The sun keeps Earth warm. Because Earth is round, the sun doesn't warm all parts of Earth evenly. That makes some places warmer than others.

Wind is caused by warm air rising and cold air taking its place. When warm air and cold air move, it makes wind. Wind can make the weather change. It can cause a rainy day to become a sunny day.

9

Wind can be light, which is called a breeze. Stronger winds can cause parts of trees to break or whole trees to be ripped out of the ground! A hurricane is a storm with very strong winds that forms over the ocean.

11

Clouds

Clouds are made up of tiny drops of water. These drops of water are so light and small that they're able to float in the air. There are three main types of clouds that bring different kinds of weather.

13

Cumulus clouds are big, **puffy** clouds that can be storm clouds. Cirrus clouds are thin clouds that often mean good weather. Stratus clouds are low, large clouds that cover most of the sky. They often bring rain or snow.

CUMULUS CLOUDS

CIRRUS CLOUDS

STRATUS CLOUDS

15

When the water drops in clouds become too heavy, they fall to the ground. If it's warm enough, these drops fall as rain. When the water in the cloud meets rising warm air, it can form a **thunderstorm**.

17

If it's colder in the cloud, then the drops of water stick together to form small pieces of ice. This ice falls to the ground as snow. Each snowflake is different!

Adapting to Weather

Animals **adapt** to the weather where they live. When the weather is cold, animals travel to warmer places or eat more to gain weight and keep warm. Many animals keep warm by not moving around.

21

Adults and kids also adapt to the weather where they live. They may stay inside and play games with their family when it rains or snows. Sometimes they put on coats and go outside to play!

WORDS TO KNOW

adapt: To change to live better in a certain place.

puffy: Rounded, soft, and light.

thunderstorm: A storm with flashes of light called lightning and loud sounds called thunder.

INDEX